INTRODUCTION.

In order to facilitate the use of the key the following definitions and notes may be of service.

The growing points together with their leaf and flower fundaments and protective coverings are termed buds The growing point representing the apex of last season's shoot and which will if uninjured continue that same axis during the succeeding season is termed a terminal bud. Buds produced elsewhere on the shoot are lateral buds, and are practically always axillary, *i.e.*, borne in the upper angle between a leaf and the stem. Buds are either scaly when provided with scales or naked when protected simply by hair or cork without scales.

The leaves when they fall leave leaf-scars, usually covered with a corky layer. The vascular bundles running through the petiole into the stem leave bundle-scars on the surface of the leaf-scars. The stipules, two little leaf-like lobes at the base of the petiole in the leaves of many species, leave stipule scars when they fall. These are always lateral to the leaf-scars one on either side. Stipules are present in only a portion of the woody plants, and in many of these they are so united with the petiole as not to leave scars distinct from the leaf-scars. Bud-scales falling away leave scale-scars ; all those from one bud forming together a bud-scar. These are most commonly met with in definite growers.

While the epidermis is still present the twigs are either hairy, or polished and smooth except for the lenticels,—little white specks serving as breathing pores. Later after cork-formation has progressed and the epidermis is thrown off the glossy appearance is lost.

Woody plants are definite or indefinite in manner of growth When definite, growth ceases usually by the middle of July after which a terminal or upper-axillary bud is formed from which the shoot of the following season is produced. When indefinite, growth continues until very late in the season. Indefinite growing shoots are usually killed back a certain distance each winter because the growing point and last few internodes have not had time to pass into a resting condition before cold weather comes.

Woody plants in winter may be recognized by bark characters, by habit characters, or by twig characters. The first two do not readily admit of description and use in a key, and vary considerably with the age of the plant. The last are much more definite, and are the ones used almost exclusively in the present key.

In the key all organs occupying the morphological position of leaves on the year's shoot, except the bud-scales are considered leaves, whether foliage-like, scale-like, or spiny. If these remain during the winter the leaves are persistent.

Care must be used in determining the presence or absence of the terminal bud. The apex of the shoot often dies regularly in certain species leaving a minute scar, while the uppermost axillary bud is pushed over so as to appear terminal. The little scar with its ring of vascular tissue, and the axillary position of the bud will enable one to detect this condition. In case the end of the year's growth regularly dies back some distance, such species have been classed also with the terminal bud absent.

When several axillary buds are present in a vertical row one above the other, they are said to be superposed.

The pith of most plants is homogeneous or hollow. When hollow the cavity may pass through the node or may be interrupted by a woody partition at that place. Pith may show transverse diaphragms of two sorts. In one case woody partitions pass through the solid pith ; in the other the pith is composed of very thin and very numerous lamellae between which are lenticular chambers,—the diaphragmed-chambered pith.

Aromatic or fragrant-aromatic is used to denote a pleasant spicy odor or taste. The term strong scented refers to rank, disagreeable or mephitic odors.

With the key and moderately good specimens in hand it is hoped that students may trace to the genus any woody plant growing either wild or cultivated within the limits of New York State. It is possible, however, that some rarer genera grown in the warmer regions around New York City have been omitted. In the preparation of the key material of only a few species of some genera was available. Where this is so the names of the species studied are inserted in parenthesis. On the other hand many genera are included which will be met with very rarely, or possibly not at all within the State. If present they will occur most likely in the parks of the larger cities or in large nurseries.

In the neighboring States the woody flora will scarcely differ from that of New York State, and hence the key will be applicable, in the main, to those regions also.

The present key has been several years in preparation, and during that time has been revised and rewritten at least six times ; still, owing to the great variability of the species and the large number of species and genera treated, the authors feel that some errors will doubtless still be found. Considering the amount of material used, however, it seems as though these must necessarily be few.

In the key one hundred and eighty-two genera are inserted from which about five hundred species were studied. With the exception of Schneider's key, ours seems to be the only one including cultivated forms. These have immensely increased the size of the key as well as the difficulty of preparation. It is hoped at a later date to present keys to the species in each genus.

LITERATURE.

Some of the most important works and papers arranged somewhat with reference to importance in this subject:

SCHNEIDER, CAMILLO KARL.—Dendrologische Winterstudien.—*Book 290 pp., 224 fig. Jena, 1903.* Very extensive; descriptions and figures of 434 species in 235 genera, cult. and native, keys to species, introductory text.

HUNTINGTON, MISS A. L.—Studies of trees in winter.—*Book, 198 pp., Boston, 1902.* Lengthy descriptions and habit notes: no key: plates.

HITCHCOCK—I. Key to Woody Plants of Manhattan (Kan.) in Winter Condition.

II. Key to the Woody Plants of Kansas.

DAVIS, K. C.—Key to the Woody Plants of Mower County in Southern Minnesota in Winter Condition, 1895.

SCHWARZ, F.—Forstliche Botanik, pp. 427–461, Berlin, 1892. Short descriptions and a partial key.

BRENDEL, FREDERICK—The Tree in Winter.—*Bull. Ill. Mus. of Nat. Hist. 1, pp. 26–32, 1876.* Points out different characters that may be used and their value; no keys nor descriptions.

FOERSTE, A. F.—The Identification of Trees in Winter.—*Bot. Gaz. 17, pp. 180–190, 1892.* Same nature as the last.

TRELEASE, W.—Has treated the Juglandaceae and Maples in the Rep. of Missouri Bot. Gard. Keys and descriptions.

EXPLANATION OF SIGNS USED.

C Cultivated.

W Native or spontaneous.

T Possibly hardy as far north as New York City.

R Rare in cultivation within the state; probably only in city parks, large nurseries or large private collections.

M Found only on alpine summits of high mountains.

CORNELL UNIVERSITY, March, 1904.

A KEY TO THE GENERA OF WOODY PLANTS IN WINTER.

A Leaves persistent throughout the winter.

b Leaves of 3–5-branched spines only ; shrubs------(C W) BERBERIS
bb Leaves spiny-margined.
  c Leaves pinnate ; shrubs-------- ------(C) (MAHONIA) BERBERIS
  cc Leaves simple ; shrubs or trees -------- -----------(C W) ILEX
bbb Leaves not spiny, though often pungent.
  c Leaves narrowly linear, subulate or appressed.
    d Leaves all either opposite or whorled.
      e Leaves of two forms, the dorsal and ventral broader and more abrupt at apex ; cone-scales basifixed ; trees.
        f Each cone-scale with 4–5 ovules----------(C T) THUJOPSIS
        ff Each scale with 2 (1–3) ovules.
          g Branches of the spray very flat ; upper two pairs of cone-scales fruitful ; leaves with dorsal papilla-(C W) THUJA
          gg Branches of spray terete ; upper pair of scales only fruitful------------ ------------(C T) LIBOCEDRUS
      ee Leaves nearly or quite similar.
        f Fruit a berry ; leaves subulate more or less appressed, or else very pungent spreading and scattered ; shrubs and trees ------------------------------------(C W) JUNIPERUS
        ff Fruit a cone with peltate scales ; leaves less pungent ; trees.
          g Cone-scale many seeded ---- ---- ------(C T) CUPRESSUS
          gg Cone-scale 2-seeded-- ----- ---- ------(C) CHAMAECYPARIS
    dd Leaves all in many small fascicles of 5 or less, each surrounded by a sheath ; trees------------------------- -- ----(C W) PINUS
    ddd Leaves alternate, scattered, terminally clustered, or some of them whorled.
      e Leaves 1– 2.5 cm. long, densely clustered at ends of short lateral branches ; trees ---------- ---- ------ (C T) CEDRUS
      ee Leaves 8–12 cm. long, whorled, also subulate scattered ones ; tree------------------------------------(C R) SCIADOPITYS
      eee Leaves evenly distributed on the branches.
        f Leaves obtuse (except rarely on leading shoots of *Pseudotsuga*, *Tsuga* and *Abies*.)
          g Low or prostrate shrubs 6 dm. high or less.
            h Leaves very small, 2–4 mm. long ; plant moss-like ; fruit a capsule--------(W M) CASSIOPE (*hypnoides*)
            hh Leaves larger, 4–7 mm. long ; plant depressed or erect ; fruit a drupe.
              i Leaves deeply furrowed beneath ; fruit terminal.
(W) COREMA

ii Leaves with prominent mid-rib beneath ; fruit axillary..........................(W M) EMPETRUM

gg Erect, more than 6 dm. high ; trees.

h Buds very resinous, often obscuring the scales, obtuse ; cones erect..........................(C W) ABIES

hh Buds not externally resinous ; cones pendent.

i Leaves 2-4 cm. long ; buds on vigorous branches large, 1 cm long, lanceolate, very acute. (C) PSEUDOTSUGA

ii Leaves 5-18 mm. long ; buds minute, 1-5 mm. long, broadly ovate, mostly obtuse........(C W) TSUGA

ff Leaves very acute or pungent (except rarely those near the fruit in *Sequoia.*)

g Leaves subulate.

h Leaves pubescent, spreading, often clustered and sharply acuminate ; shrub..............(C T) ULEX

hh Leaves glabrous.

i Leaves long-subulate 1-1.5 cm. long, rigid, 4-angled, falcate......................(C T) CRYPTOMERIA

ii Leaves broad-subulate, minute, scale-like, acuminate..............................(C) TAMARIX

gg Leaves linear (subulate near the fruit in *Sequoia.*)

h Leaves tetragonal..........................(C W) PICEA

hh Leaves flat.

i Buds very resinous, often obscuring the scales ; glaucous stomate-bearing lines on the underside of leaves broader than the green bands..(C W) ABIES

ii Buds not resinous on the surface.

j Leaves with the 2 white or browish stomate-bearing lines beneath narrower than the 3 green bands..........................(C T) TORREYA

jj Leaves with stomate-bearing bands broader than the 3 green bands.

k Leaves petioled..................(C W) TAXUS

kk Leaves sessile, decurrent.......(C R) SEQUOIA

cc Leaves elliptic-linear, or broader, spreading ; shrubs.

d Leaves opposite.

e Leaves crenate ; upright, climbing or prostrate shrubs.

f Branches green, glabrous, often 4-angled ; erect, prostrate or climbing shrubs. (C R) EUONYMUS (*Japonicus radicans*)

ff Branches brownish, pubescent when young, terete ; plant creeping..............................(W) LINNAEA

ee Leaves sharply serrate ; plant low, .5- 1.5 dm. high, erect. (W) CHIMAPHILA

eee Leaves entire.

f Plant creeping..........................(W) MITCHELLA

ff Plant erect.

g Margins of petioles strongly decurrent; veins very close together, pinnately-parallel .... (C) Buxus

gg Margins of petioles not decurrent; veins distant, forming broad areoles.

h Leaves sessile and twigs 2-edged; or petioled and verticillate with terete twigs and naked buds (W) Kalmia (*glauca*, *augustifolia*)

hh Leaves petioled, opposite, usually falling at mid-winter or before; twigs terete; buds ovate acutish, scaly............................ (C) Ligustrum

dd Leaves alternate.

e Climbing on walls, rocks, trees, etc. with root-like hold-fasts, leaves usually 3-5-lobed .... (C) Hedera (*Helix*)

ee Creeping, prostrate, or forming dense cushion-like tufts.

f Leaves large, oval-oblong, 2.5- 7.5 cm. long, rugose. (W) Epigaea

ff Leaves much smaller or else not oval-oblong and rugose.

g Leaves mucronate.

h Leaves with scattered brown scale-like hairs beneath and also on stem; leaves 12 mm. long or less. (W) Chiogenes

hh Leaves with no scattered hairs as above, 14- 40 mm. long...(C R) Daphne (*cneorum* and *Blagayana*)

gg Leaves obtuse or acute, not mucronate; no brown hairs beneath.

h Stems densely tufted, very short; leaves spatulate often recurved; alpine................ (W M) Diapensia

hh Stems not densely tufted; long and slender.

i Leaves white beneath; bog plants..(W) Oxycoccus

ii Leaves green beneath.

j Leaves black-dotted beneath, 6- 16 mm. long; alpine...................... (W M) Vitis Idaea

jj Leaves not black-dotted.

k Leaves 15- 30 mm. long, obovate-spatulate (C W) Arctostaphylos (*Uva-ursi*)

kk Leaves 2-7 mm. long.

l Leaves 2-4 mm. long; plant moss like; fruit a capsule...... (W M) Cassiope (*hypnoides*)

ll Leaves 4-7 mm. long; plant depressed or prostrate; fruit a drupe ...(W M) Empetrum

eee Erect.

f Thorny.............................. (C T) Pyracantha

ff Not thorny.

g Leaves 4-7 mm. long, deeply furrowed beneath. (W) Corema

gg Leaves larger, not furrowed beneath.

h Low, 1-2 dm. high, from subterranean stems; leaves oval (W) Gaultheria (*procumbens*) *hh* Higher 3 dm. or more.

i Leaves densely pubescent or tomentose beneath.

j Leaves 5–15 cm. long--------(C) RHODODENDRON

jj Leaves .25– 5.00 cm. long.

k Tomentum white ; leaves not rugose.
(C R) COTONEASTER

kk Tomentum brown ; leaves rugose.
(W) LEDUM (*Groenlandicum*)

ii Leaves nearly or quite without hairs.

j Leaves scurfy.

k Twigs about 1 mm. diam.; uppermost leaves smaller and mostly secund.
(W) CHAMAEDAPHNE

kk Twigs stouter ; uppermost leaves not conspicuously as above -------(C) RHODODENDRON

jj Leaves not scurfy.

k Leaves crenulate, usually strigose beneath and ciliate, rugose ; twigs coarsely strigose.
(C R) PIERIS (*floribunda*)

kk Leaves not crenulate, or if so then twigs not coarsely strigose.

l Leaves and branches more or less strigose.
(C T) AZALEA (*Indica*)

ll Leaves glabrous ; branches rarely pubescent.

m Leaves lance-linear, very white beneath.
(W) ANDROMEDA (*polifolia*)

mm Leaves usually broader, if lance-linear not white beneath.

n Leaves scattered on the year's growth, teeth blunt-----(C T) ILEX (*glabra, crenata, vomitoria.*)

nn Leaves clustered at end of year's growth, entire.

o Years growth with numerous scale-scars scattered below the leaves ; latter rugulose above, whitish beneath ; petiole at juncture with stem spreading ---(C W) RHODODENDRON (*maximum, Catawbiense*, etc.)

oo Year's growth with only 2 scale scars at very base ; leaves not rugulose, light green beneath ; petiole at juncture with stem erect ----(C W) KALMIA (*latifolia.*)

AA Leaves Deciduous.

b Leaf-scars absent ; twigs deciduous leaving twig-scars ; Gymnosperms, wood without vessels ; trees-----------(C T) TAXODIUM *bb* Leaf-scars except on young shoots mainly densely clustered on short branches (see also *Betula*); bark often full of resin; wood without vessels; Gymnosperms.

c Scattered leaf-scars also present on twigs, very numerous and strongly decurrent; twigs and branches slender, brownish; trees ....(C W) LARIX

cc Scattered leaf-scars if present large, not decurrent, and separated 1.5 cm. or more; twigs and branches stout, light brown; bundle-scars mostly 2; upper edge of leaf-scar fringed; tree. (C) GINKGO

bbb Leaf-scars opposite or whorled; wood with vessels; Angiosperms.

c Twigs and buds densely peltate-scaly; shrubs. (C W) LEPARGYRAEA

cc Twigs and buds not peltate-scaly.

d Stipule-scars 4 at each node, lunate-triangular, distinct from leaf-scar; bark finely white-striped; shrubs. (C W) STAPHYLEA

dd Stipule-scars absent or represented by a line connecting the leaf-bases.

e Stems twining or climbing.

f Pith with woody partitions at the nodes.

g Petioles remaining as tendrils; pith solid. (C W) ATRAGENE

gg Petioles not tendril-like.

h Bundle-scars 3..............................(C W) LONICERA

hh Bundle-scar 1; bud hidden behind projecting leaf-scar.......................................(C) PERIPLOCA

ff Pith continuous at the nodes; no tendrils, but root-like holdfasts.......................................(C) TECOMA

ee Stems erect.

f Bundle-scars numerous, arranged in a longitudinal closed ellipse; leaf-scars orbicular or longitudinal oval, often whorled; terminal bud absent; trees....(C W) CATALPA

ff Bundle-scars one, or very numerous and almost confluent (often forming a transverse lunate or U-shaped line.)

g Buds superposed but one often minute (rarely also in one sp. *Syringa*.)

h Buds nearly or quite glabrous, of medium size.

i Buds oblong; at least upper leaf scars decurrent; twigs glabrous; shrubs............(C) FORSYTHIA

ii Buds short ovate; leaf scars scarcely decurrent; twigs often hirsute; shrubs.... (C) CHIONANTHUS

hh Buds tomentose, medium or small.

i Leaf-scars strongly V-shaped; twigs strongly aromatic; buds usually distant; shrub. (C R) VITEX (*Agnus-Castus*)

ii Leaf-scars not V-shaped; twigs scarcely aromatic; buds usually contiguous.

j Buds partly unfolded in winter; hairs of stem slender, thinly tomentose; shrub.
(C T) CARYOPTERIS

jj Buds not unfolded, minute; hairs short, scurfy-scabrous; shrub.(C R) CALLICARPA (*purpurea*)

gg Buds solitary in the axils or lateral, or accompanied by a branch of the previous season.

h Leaf-scars strongly decurrent.

i Twigs bright green.

j Leaves deciduous above the orange-brown base; scar ragged; shrub.
(C T) JASMINUM (*nudiflorum*)

jj Leaves deciduous at the green base; scar clear-cut; branches often winged; shrub.
(C R) EUONYMUS (*alatus*)

ii Twigs brown or olive.

j Buds tomentose; branches mostly membraneous-winged or quadrangular; shrub.
(C R) BUDDLEIA (*Japonica*)

jj Buds glabrous; branches not winged nor distinctly 4-angled.

k Buds minute 1–2 mm. long; shrub.
(C R) FONTANESIA (*Fortunei*)

kk Buds larger, 2–12 mm. long; leaf scars but slightly decurrent; shrub or small tree.
(C) SYRINGA (*Persica*)

hh Leaf-scars scarcely or not at all decurrent, though twigs sometimes with corky angles.

i Buds naked, partly opened in winter, tomentose; shrubs ---------------------(C T) CARYOPTERIS

ii Bud-scales very loose, very acute, at first herbaceous, finally withering and drying; withered leaves persistent; twigs glabrous, bronze-brown; shrubs.
(C) HYPERICUM (*Moserianum*)

iii Bud-scales, mostly close, normal.

j Leaf-scars connected by a stipular line

k Buds submerged in bark; twigs medium or stout; shrub or small tree.
(C W) CEPHALANTHUS

kk Buds normal axillary; twigs slender; low shrubs-------------(C W) SYMPHORICARPOS

jj Leaf-scars not connected by a stipular line (indistinct one sometimes in *Fraxinus*).

k Twigs bright green, 4-angled, except the youngest; shrub or tree---------(C W) EUONYMUS

kk Twigs brown, gray, or rarely olive.

l Buds appressed, black, acute; shrub.
(C W) RHAMNUS

ll Buds divaricate or buds very obtuse and pale.

m Twigs 1.5 mm, or less diam., hairy or smooth; term. bud often 4-angled, its scales acute; shrubs--(C W) LIGUSTRUM

mm Twigs 2 mm. or more diam.

n Buds scurfy, brown or black; bundle scars often almost separate and very numerous, forming a long U-shaped line; trees------------------(C W) FRAXINUS

nn Buds not scurfy, nearly or quite glabrous, paler, bundle-scar lunate.

o Buds minute 2mm. long very obtuse, not angled; twigs usually hirsute, shrub. (C) CHIONANTHUS

oo Buds larger, 3mm. or more long, acutish, more or less 2–4 angled; twigs mostly glabrous; shrub or small tree. (C W) SYRINGA (*vulgaris*)

fff Bundle-scars 3 or more in a lunate or U-shaped line, when more than three quite distinct from each other (see also young canes of *Symphoricarpos racemosus*).

g Twigs fragrant-aromatic; shrubs (C) BUTNERIA (*florida*)

gg Twigs not fragrant-aromatic; buds sometimes superposed.

h Buds bursting through the leaf-scars; shrubs. (C) PHILADELPHUS

hh Buds axillary.

i Stem hollow except in young twigs, then nearly so.

j Leaf-scars mostly flat; twigs mostly brown or bronze, usually with stellate hairs; buds not superposed; shrubs------ ----------(C) DEUTZIA

jj Leaf-scars with upper margin acutely projecting, or buds superposed; twigs gray or whitish; no stellate hairs; shrubs -----------(C) LONICERA

ii Stem solid.

j Bud-scales of axillary buds 0–3 pairs (sometimes with 1–2 pairs of extra bracteoles beneath).

k Buds without scales, densely tomentose; the pinnately lobed foliage leaves serving as scales; shrubs. (C W) VIBURNUM (*alnifolium, lantana*)

kk Buds scaly, not tomentose often silky (scales becoming petioles in *Viburnum Lentago*).

l Buds scurfy, long linear-lanceolate, often curved; shrubs. (C W) VIBURNUM (*Lentago, prunifolium*)

ll Buds not scurfy, shorter.

m First pair of scales shorter than the bud; shrubs.
(C W) VIBURNUM (*dentatum*, *acerifolium*)

mm First pair of scales equalling the bud (until swelling begins).

n Line connecting uppermost pair of leaf-scars notched; shrubs---(C W) CORNUS

nn Line not notched.

o Second pair of bud-scales hairy; shrubs or trees.
(C W) ACER (*Pennsylvanicum*, *spicatum*)

oo Second pair of bud-scales glabrous or glutinous; shrubs.
(C W) VIBURNUM (*Opulus*)

jj Bud-scales of axillary buds 4–many pairs.

k Bundle-scars normally 5–many; twigs stout.

l Buds depressed, conical and acute, or oval and green or purple; pith large; shrubs.
(C W) SAMBUCUS (*Canadensis*, *pubens*)

ll Buds large, ovate-oblong, gray-brown or varnished; trees-------------(C W) ÆSCULUS

kk Bundle-scars normally 3, rarely in some scars 5; twigs stout or slender.

l Decurrent ridge from middle of line connecting leaf-scars.

m Upper edge of leaf-scar straight or convex; middle of scar slightly or not at all decurrent; shrub-------(C W) DIERVILLA

mm Upper edge of leaf-scar concave; lower angle sharp, projecting and strongly decurrent; shrub ----(C R) LONICERA (*involucrata.*)

ll No decurrent ridge as above.

m Buds clustered in the axils, loose—scaly; shrubs--------------(C R) RHODOTYPOS

mm Buds solitary in the axils or superposed.

n Upper edge of leaf scars nearly straight.

o Buds depressed, appearing as though just breaking through the epidermis; or bud-scales of previous year's bud persistent; shrubs or small trees--(C) HYDRANGEA

oo Buds exserted and normal, and bud scales deciduous.

p No line connecting leaf-scars; shrubs
(C W) RHAMNUS

pp Line connecting leaf-scars evident; shrubs----------(C W) LONICERA *nn* Upper edge of leaf-scars strongly concave.

o Buds scurfy; twigs moderately stout; shrub --------(C) AESCULUS (*Pavia*)

oo Buds not scurfy; twigs stout or slender.

p Line joining leaf-scars strongly ligulate; buds whitish silky; twigs olive usually glaucous; trees------(C W) ACER (*Negundo.*)

pp Line mucronate or straight; trees. (C W) ACER.

bbbb Leaf-scars alternate; wood with vessels; Angiosperms.

c Bundles in stem scattered; brier-like, green, often prickly shrubs; Monocotyledons--------------------------------(W) SMILAX

cc Bundles in stem in a ring; Dicotyledons.

d Stem climbing or twining.

e With tendrils.

f Woody partitions through the brownish pith at the nodes. (C W) VITIS

ff No woody partitions; pith continuous, white. (C W) (AMPELOPSIS) PARTHENOCISSUS

ee Without tendrils.

f Pith diaphragmed-chambered; buds sunk in the cortex; bundle-scar one--------------------------(C) ACTINIDIA

ff Pith continuous or hollow.

g Buds in slightly superaxillary depressions; leaf-scars usually orbicular with several bundles and with or without a narrow sinus; twigs green. (C W) MENISPERMUM (*Canadense*)

gg Buds normal, axillary; leaf-scars lunate, semi-circular, V-shaped or orbicular.

h Stems prickly; leaf-scars very narrow------ (C) ROSA

hh Stems not prickly.

i Bundle-scars 3–5, distant.

j Buds tomentose-silky.

k Twigs green; buds clustered-(C) ARISTOLOCHIA

kk Twigs gray; buds solitary; juice resinous. (W) RHUS (*radicans*)

jj Buds glabrous; leaves deciduous slightly above the base-------------------- (C) AKEBIA (*quinata*)

ii Bundle-scar one, or several and confluent.

j Stem hollow; twigs angled, green or whitish, with strong odor--------(W) SOLANUM (*Dulcamara*)

jj Stem solid; odor not strong.

k Buds silky; leaf-scars often two-horned at base, projecting, not decurrent. (C) (WISTARIA) BRADLEYA (*frutescens*)

kk Buds glabrous, or very nearly so.

l Leaves not decurrent; buds pungent, solitary; twigs gray....(C W) CELASTRUS (*scandens*)

ll Leaves decurrent; buds not pungent, often clustered; twigs silvery-whitish, often thorny. (C W) LYCIUM (*vulgare*)

dd Stems not twining nor climbing, usually erect.

e Bundle-scars more than one.

f Bundle-scars more than 3, in a closed ellipse, double line, irregularly scattered, or clustered.

g Stipule-scars or stipules present.

h Terminal bud present (persistent stipules sometimes present in *Opulaster*.)

i Stipule-scars extending one-half way around stem, or more.

j Bud-scales many; buds lanceolate, acute; trees. (C W) FAGUS

jj Bud-scales 2, united; buds scarcely acute; twigs usually aromatic.

k Buds hairy; leaf-scars mostly lunate; trees. (C W) MAGNOLIA

kk Buds glabrous; leaf-scars mostly orbicular; trees..................(C W) LIRIODENDRON

ii Stipule-scars not extending half way around the stem.

j Buds depressed, indistinct, white tomentose with branched hairs; scales irregular; shrubs or small trees.........................(C) HIBISCUS

jj Buds ovate, prominent, not tomentose, mostly clustered at ends of twigs; scales closely imbricated; trees...................(C W) QUERCUS

hh Terminal bud absent.

i Buds depressed, flat-topped, usually lateral to axillary thorns; branches light gray-olive; bundle scars practically confluent; shrubs or trees. (C) TOXYLON

ii Buds ordinary; twigs never thorny.

j Uppermost bud broader than the subtending internode (rarely also in sp. of *Morus* with papery scales).

k Visible bud-scales 4–many, usually hairy; catkins usually present; twigs often hairy; shrubs.......................(C W) CORYLUS

kk Visible bud scales 2–3, glabrous or nearly so; no catkins; twigs glabrous.

l Twigs olive bronze, often angled, nearly straight; buds often rugose; trees. (C W) CASTANEA *ll* Twigs red, yellow or green, usually zigzag ; trees........................(C W) TILIA

jj Uppermost bud as broad as or narrower than the subtending internode.

k Twigs scabrous, often mottled ; buds acute or acuminate with 2–3 visible scales ; trees. (C R W) BROUSSONETIA

kk Twigs not scabrous, not mottled ; buds acute or obtuse, with many visible scales ; trees. (C W) MORUS

gg Stipule-scars and stipules absent. (See also *Toxylon*).

h Terminal bud present (flower bud in *Rhus*).

i Leaf-scars nearly circular, 1–2 mm. broad ; twigs slender ; beginning of years growth abruptly constricted ; axillary buds not visible ; odor strong, disagreeable ; low or prostrate shrub. (C W) RHUS (*Canadensis*)

ii Leaf scars inversely triangular or oblong ; twigs stout.

j Bark of twigs mottled, juice resinous ; buds small ; shrub ....(W) RHUS (*vernix*)

jj Bark of twigs not mottled, often lenticellate, juice not resinous ; buds very large ; trees. (C W) HICORIA

hh Terminal bud absent ; buds laterally compressed, with 2 lateral scales ; upper edge of leaf-scar projecting ; trees...............(C R) KOELREUTERIA

ff Bundle-scars 3 or more in a single lunate line.

g Stipule-scars or stipules present (spines or prickles are not here considered stipules.)

h Terminal bud present.

i Buds pedicelled, more or less unsymmetrical.

j Buds densely tomentose ; tall shrub. (C W) HAMAMELIS (*Virginiana*)

jj Buds scurfy ; shrub or tree.........(C W) ALNUS

ii Buds sessile nearly symmetrical.

j First scale of axillary bud anterior.(C W) POPULUS

jj First scales of axillary bud lateral.

k Twigs brier-like, often prickly ; leaves deciduous several mm. above the base, scars very rough and uneven...................(C W) RUBUS

kk Twigs not brier-like, never prickly ; leaves falling at or near the base, scar smoother.

l Stipules persistent firm, subulate ; shrub. (C R) CARAGANA

ll Stipules deciduous.

m Leaf-scars projecting uneven ; low shrub with weak olive-brown twigs.
(C W) CEANOTHUS (*Americanus*)

mm Leaf-scars not conspicuously projecting, even ; twigs very firm, often thorny (see also Betula with wintergreen flavored bark.)

n Twigs light olive and dull, or twigs hirsute ; leaf scar usually narrowly lunate.
(C) CYDONIA (*Jap., vulg.*)

nn Twigs olive reddish or darker, polished or covered with white crust, glabrous ; leaf-scars usually broadly lunate.
(C W) PRUNUS (*peach, cherry, mostly*)

hh Terminal bud absent.

i Buds superposed.

j Exposed bud scales 2–3 ; twigs, gray-brown, striate ; shrubs............(C R) AMORPHA (*fruticosa*)

jj Exposed bud-scales many ; twigs light brown, not striate ; shrubs..........(C R) STEPHANANDRA

ii Buds not superposed.

j Stipule-scar extending entirely around the stem ; trees ....(C W) PLATANUS (*occidentalis*)

jj Stipule-scars not extending around the stem.

k Bud scale one ; shrubs and trees ..(C W) SALIX

kk Bud scales several, or buds sunk in leaf-scar.

l Buds depressed in leaf-scar ; densely brown-felty ; twigs green ; trees ...(C R) SOPHORA (*Japonica*)

ll Buds not depressed, not densely brown-felty, but sometimes strigose ; twigs various.

m Visible bud scales 1–3 (rarely 4).

n Bud-tips appressed, brown ; bundle-scars almost confluent ; pith usually diaphragmed-chambered ; tree .....(C W) CELTIS (*occidentalis*)

nn Bud-tips not appressed ; pith not diaphragmed.

o Buds globular ovate, glabrous, broader than the usually zigzag twigs ; tree.
(C W) TILIA.

oo Buds ovate-lanceolate, acute, slightly broader than the twig ; latter not zigzag ; bark often with wintergreen flavor ; buds on leafy spurs terminal ; trees................(C W) BETULA.

mm Visible bud-scales 4-many.

n Leaf-scars in one-half phyllotaxy.
o Year's growth densely clothed at base with leaf-scars ; numerous short spurs on older wood densely clothed with leaf-scars and possessing a terminal bud ; bark often aromatic ; trees.
(C W) BETULA.
oo Years growth not as above, and no such spurs.
p Leaf-scars 1.75- 4 mm. diam., covered with a smooth corky layer in which bundle-scars are depressed ; buds 4-8 mm. long ; spray rather coarse 1.25 mm. or more diam. ; trees.
(C W) ULMUS.
pp Leaf-scars 1.75 mm. or less diam., not smooth and corky ; buds 7 mm. or less long ; spray slender 1.75 mm. or less diam.
q Buds usually 3-7 mm. long ; spray 1.5-1.75 mm. thick ; bark of trunk very flaky and rough, dark gray ; tree ------ (C W) OSTRYA
qq Buds usually 2-4 mm. long ; spray about 1 mm. thick ; bark of trunk smooth furrowed light gray ; shrub or low tree.
(C W) CARPINUS (*Caroliniana*).
nn Leaf scars not in one-half phyllotaxy.
o Briar-like plants, often prickly ; leaves deciduous several mm. above base ; shrub------------------ C W) RUBUS
oo Not briar-like, never prickly ; leaves not deciduous above the base.
p Bark of whole plant bright green ; leaves decurrent ; shrubs.
(C). KERRIA
pp Bark of older branches not green.
q Buds ovate ; twigs not glandular.
r Bud-scales dark brown or black ; twigs gray or green ; shrubs.
(C W) RHAMNUS
rr Bud scales red or light brown ; twigs olive or red brown ; shrub, (C) CYDONIA (*Japonica*)
qq Buds globular ; twigs glandular, dark ; shrubs.
(C W) COMPTONIA

gg Stipule-scars and stipules absent.

h Terminal bud present.

i Pith diaphragmed-chambered with very thin lamellae; buds superposed, with thick irregular scales; trees, (*Pterocarya* may possibly occur in the State). (C W) JUGLANS

ii Pith with transverse woody partitions through the solid ground mass.

j Bundle-scars 5–7; buds globular, densely velvety with long dark hairs; tree ...... (C W) ASIMINA

jj Bundle-scars 3; buds ovate, nearly or quite glabrous; trees .... (C W) NYSSA

iii Pith homogeneous.

j Bundle-scars 5–30.

k Stems prickly.

l Leaf-scar not extending more than one-half way around stem; bundle-scars 5–10; shrubs or trees .... (C R) ACANTHOPANAX (*pentaphyllum*, *riciniphyllum*)

ll Leaf-scar extending nearly around the stem; bundle-scars about 20; shrubs or trees. (C) ARALIA (*spinosa*)

kk Stems not prickly.

l Wood bright yellow; leaf scars extending almost around the stem; low shrubs. (C W) XANTHORHIZA

ll Wood not bright yellow; leaf-scars not extending more than half way around stem.

m Pith 1–2 mm. diam.; bark without sticky juice; buds dark red, hairy, mostly appressed; tree. (C W) SORBUS (*Aucuparia*, *Americana*)

mm Pith 1.75–5 mm. diam.; shrubs.

n Bark with sticky juice; bud-scales of the medium-sized buds few. (W) RHUS *radicans*, *vernix*)

nn Bark without sticky juice; buds large, the scales 10 or more. (C R) PAEONIA (*Moutan*)

jj Bundle-scars 3

k Leaf-scars very narrow, a mere line extending one-half way or more around stem, not decurrent; shrubs, often brier-like, prickly or smooth .... (C W) ROSA

kk Leaf-scars broader, often decurrent.

l Twigs prickly, or buds red-tomentose.

m Leaf deciduous above base; twigs recurved, brier-like; shrubs .... (C W) RUBUS *mm* Leaf deciduous at base; twigs not conspicously brier-like.

n Prickles 1–3 below each leaf-scar, the latter decurrent; twigs pale and buds lanceolate with thin scales; shrub.
(C W) RIBES (*oxyacanthoides*)

nn Prickles 2 at each leaf scar or scattered, scar not decurrent; buds depressed, in one sp. red tomentose; shrub or small tree-----------(C W) ZANTHOXYLUM

ll Twigs not prickly, sometimes thorny, and buds not red tomentose.

m Juice of bark resinous and strong scented; leaf-scars semicircular or broadly lunate; shrub or tree-------------(C) COTINUS

mm Juice of bark not resinous, not strong scented, often fragrant-aromatic.

n Leaf-scars semi-circular or broadly lunate, large, 2.5 mm or more diameter.

o First bud scale nearly as long as the bud; axillary buds deltoid, flattened; twigs bronze; shrubs or trees.
(C) XANTHOCEROS

oo First bud scale one-half length of bud or less.

p Older branches corky-angled, dark ash-colored; twigs and buds reddish olive; bundle-scars very large, annular; trees-- (C W) LIQUIDAMBAR

pp Older branches not corky-angled, paler; bundle scars not annular; trees.
(C W) POPULUS (*grandidentata*)

nn Leaf-scars lunate or else quite small and 2 mm. or less diameter.

o Internodes very unequal, one each year many times the length of the rest except on ultimate branchlets; branches much exceeding the parent axis; tall shrubs--(C W) CORNUS (*alternifolia*)

oo Internodes not very unequal; branches usually shorter than the parent axis.

p Leaf scars strongly decurrent from sides and middle, very small 1.5 mm. diam., somewhat triangular; twigs bronze-brown; bundle-scars contiguous; shrub.
(C W) OPULASTER (*opulifolius*)

pp Leaf-scars not decurrent; or if so then leaf scars lunate and longer, 2.25 mm. diam., and bundle scars widely separated.

q Buds superposed: bark aromatic; shrub............ (C W) BENZOIN

qq Buds not superposed; bark not aromatic.

r Buds elliptic, lanceolate, or lance-linear.

s At least older bark shreddy; bud scales very thin light or dark brown, often glandular; leaf-scars often decurrent; shrubs.......... (C W) RIBES

ss Bark close; buds pale brown or red; leaves not decurrent.

t Second bud scale nearly one-half length of bud or more, rather thick; shrubs. (C W) ARONIA

tt Second scale one-half length of bud or less, thinner and closely appressed; shrubs or trees. (C W) AMELANCHIER

rr Buds ovate or depressed.

s Leaf-scars strongly projecting (2–4 mm.), small (about 1.5 mm. diam.), buds long-hairy, irregular; twigs rather stout, green; shrubs. (C R) CYTISUS

ss Leaf scars not as above.

t Bark of twigs pale brown, dull and rough-fissured; buds reddish; shrub. (C R) EXOCHORDA (*grandiflora*)

tt Bark of twigs polished, shreddy or hairy, pale-gray, olive or dark.

u Terminal bud-scales narrowly ovate, thick, red, mostly 3-dentate; the bud narrowly-ovate; twigs dark-red or bronze-gray; small tree. (C W) MALUS (*coronaria*) *uu* Terminal bud-scales broadly ovate, thinner and appressed; the bud usually broadly ovate.

v Axillary buds flattened and closely appressed (rarely also appressed in *Pyrus communis*, but buds then acute, pungent), broadly ovate, mostly hairy; leaf scars rather narrow; twigs mostly dark; tree.
(C W) MALUS

vv Axillary buds plump and mostly divaricate, often pungent; twigs light or dark.

w Leaves deciduous above the base; brier-like plants; shrubs.
(C W) RUBUS

ww Leaves deciduous at the base.

x Buds obtuse, rarely acute, solitary, scales thick; leaf-scar narrow; branches often thorny; trees.
(C W) CRATAEGUS

xx Buds acute, often clustered; scales thin and appressed; no true thorns.

y Buds conical, pungent, solitary; leaf-scars narrow; twigs mostly yellow-olive; trees.
(C W) PYRUS (*communis*)

yy Buds ovate, slightly contracted below, often clustered: leaf scars broad; twigs various; trees.
(C W) PRUNUS (peach and cherry)

hh Terminal bud absent.

i Buds hidden or sunken in depressions in the bark or leaf-scar.

j Depressions superaxillary; buds tomentose; pith red or pink; trees--------(C W) GYMNOCLADUS

jj Depressions not superaxillary; pith white or yellow.

k Buds glabrous, axillary; twigs dark-olive or red, polished, often thorny; trees. (C W) GLEDITSIA

kk Buds very hairy, breaking through the leaf-scar; twigs brown, gray or viscid, duller, often prickly; trees.---------(C W) ROBINIA

ii Buds not sunken.

j Bundle-scars, 5–many.

k Leaf-scars extending nearly around the stem; twigs stout, prickly and aromatic; shrub or small tree-----------------------(C) ARALIA

kk Leaf-scars not extending half way around stem; twigs not prickly nor aromatic.

l Leaf-scars extending nearly around the bud, deeply V-shaped; buds tomentose or densely silky.

m Bundle-scars projecting out of the almost white leaf-scars; twigs dark-olive; tree. (C) CLADRASTIS (*lutea*)

mm Bundle-scars not projecting; twigs bronze-gray or pale.

n Twigs 4 mm. or more diam., very stout; bark with resinous juice; shrubs or trees. (C W) RHUS (*hirta, glabra*)

nn. Twigs 2 mm. or less diam., abruptly contracted at end of each year's growth; bark not resinous but very strong and fibrous; shrubs -------(C W) DIRCA

ll Leaf-scars semicircular deltoid or lunate, not surrounding the bud.

m Buds depressed; tree---(C W) AILANTHUS

mm Buds very large, ovate, acute; shrub. (C) PAEONIA (*Moutan*)

jj Bundle-scars, 3.

k Leaf-scars, deeply V-shaped, partly surrounding the hairy bud.

l Buds rusty-hairy, longitudinally oval; bark not fetid; twigs stout; tree. (C R) PHELLODENDRON (*Amurense*) *ll* Buds whitish-hairy, round, mostly superposed; bark with fetid odor; twigs moderate; small tree........................(C W) PTELEA

kk Leaf-scars not deeply V-shaped.

l Leaf-scar very narrow almost a line, not decurrent, extending one-half way around stem; brier-like and often prickly; shrubs. (C W) ROSA

ll Leaf-scars broader and shorter.

m Bark of whole shrub bright green. (C) KERRIA

mm Bark of older twigs not green.

n Buds mostly superposed.

o Twigs speckled and often zigzag, dark; trees..................(C W) CERCIS (*Canadensis*, *Japonica*)

oo Twigs green or gray-brown, furrowed; shrubs. (C W) BACCHARIS (*halimifolia*)

nn Buds not superposed.

o Bark with resinous juice; leaf-scars semicircular; shrub or small tree. (C W) RHUS

oo Bark not resinous; leaf-scars lunate.

p Leaves deciduous 3–4 mm. above base.

q Bud-scales blunt terminated by a scar, glabrous or nearly so; prickles if present foliar; inner bark bright yellow; shrub. (C W) BERBERIS (*vulgaris*)

qq Bud-scales not blunt and terminated by a scar; often densely silky; prickles, if present, not foliar; shrub......(C W) RUBUS

pp Leaves deciduous at the base.

q Two long prickles at each node, one shorter than the other and recurved; low shrubs.(C) PALIURUS

qq No prickles

r Twigs dull, brown, and often glandular or nearly black, slender and straight; internodes short; low shrubs of wet grounds.......... (W) MYRICA

rr Twigs polished, olive-bronze or red; internodes longer; large shrubs or trees.

s Buds depressed, obtuse, polished, with firm scales; twigs mostly zigzag.
(C W) CRATAEGUS (*tomentosa*, etc.)

ss Buds acute or obtusish, dull; scales thin: twigs straight.
(C W) PRUNUS (plums)

ee Bundle-scar one.

f Twigs peltate-scaly or stellate-pubescent.

g Young twigs stellate-pubescent.

h Stipule-scars absent; shrub.
(C W) CLETHRA (*alnifolia*)

hh Stipule scars present; shrubs--- (C R) FOTHERGILLA

gg Young twigs peltate-scaly.

h Terminal bud present; buds 2–3 mm. long, silvery or rusty; shrub or small tree.
(C) ELAEAGNUS (*argentea*)

hh Terminal bud absent; buds 2–8 mm. long, rusty; twigs often thorny; shrubs -------(C) HIPPOPHAË

ff Twigs not peltate-scaly nor stellate.

g Stipule-scars or stipules present.

h Twigs coarsely striate with green or brown and gray stripes: buds strongly superposed; shrub.
(C R) AMORPHA (*fruticosa*)

hh Twigs not markedly striate.

i Stipules deciduous.

j Leaf-scars projecting; twigs not thorny, weak, olive-green or reddish-brown; low shrubs.
(C W) CEANOTHUS (*Americanus*)

jj Leaf scars not projecting; twigs usually thorny with buds lateral to the thorns, gray or gray-green; bundle scar annular; shrub or tree.
(C) TOXYLON
(See here also *Celtis* when the bundle-scars are sometimes fused into one.)

ii Stipules persistent. (See also TOXYLON.)

j Stipules sheathing the stem; bark, deep, bright brown, shreddy; shrubs------(C W) DASIPHORA

jj Stipules not sheathing the stem; bark dull brown or gray.

k Twigs strongly and sharply, many-angled, bright green; stipules minute; shrubs.
(C T W) GENISTA (*tinctoria*)
(C T) CYTISUS (*scoparius*)

kk Twigs not as above.

l Leaf-scars not projecting; stipules minute; twigs light gray-brown; buds often superposed; fruit a red berry; shrubs.
(C W) ILEX (*verticillata*)

ll Leaf-scars projecting; bark green, olive-gray, or yellowish brown.

m Stipules deltoid; bark olive-gray; twigs glabrous; shrubs.
(C R) COLUTEA (*arborescens*)

mm Stipules firm-subulate.

n Twigs stout, about 3 mm. diam.; bark yellowish-brown; leaf-scar with 3 decurrent lines; shrub or small tree.
(C R) CARAGANA (*arborescens*)

nn Twigs slender, 1- 1.5 mm. diam.; olive-brown, glabrous; shrubs.
(C R) HALIMODENDRON

gg Stipule-scars and stipules absent. (See also *Toxylon*)

h Terminal bud present.

i Bark aromatic; twigs green; trees.
(C W) SASSAFRAS

ii Bark not aromatic.

j Leaf-scars not projecting beyond the plane of the epidermis; very smooth and even.

k Buds, except on strong shoots, mostly clustered at ends of twigs, solitary; twigs brown or ashy; terminal bud often very large; shrub.
(C W) AZALEA (*nudiflora*)
(*Rhodora* may be in N. Y. State)

kk Buds not clustered, often superposed; twigs light gray-brown, speckled; shrub.
(C W) DAPHNE (*Mezereum*)

jj Leaf-scars projecting slightly; rough and very uneven; shrubs................(C W) SPIRAEA
(*Stuartia* would come here).

jjj Leaf-scars projecting slightly, slightly uneven; buds small, all of about the same size; twigs light gray; shrubs.
(C W) (NEMOPANTHES) ILICIOIDES

hh Terminal bud absent.

i Twigs finely white-speckled and granulated, green or reddish; buds often slightly acuminate; shrub.
(C W) VACCINIUM

ii Twigs not white-speckled as above.

j Pith diaphragmed-chambered; buds superposed; tree ....(C T) MOHRODENDRON

jj Pith not as above; buds solitary.

k Buds of two sorts, large flower buds 3.3 mm. long or more with many visible scales, small short buds with 2 visible scales, reddish-yellow divaricate; shrub.

(C W) GAYLUSSACIA (*resinosa*)

kk Buds all alike.

l Buds appressed, acute; visible scales 2; second years' twigs light brown; shrub.

(C W) XOLISMA

ll Buds divaricate.

m Visible scales 2, dark-brown; twigs stout; leaf-scars large, 2.5 mm. diam.; the upper edge very sharp; bundle-scar transversely elongated; trees...(C W) DIOSPYROS

mm Visible scales several.

n Buds chocolate-brown or reddish; scales very numerous, outer usually much less than half length of bud; surface of leaf-scar very uneven; shrub.

(C W) SPIRAEA

nn Buds bronze-brown, red or olive; surface of leaf-scar not so uneven.

o Twigs stout, 1.75 mm. diam. or more, highly polished, olive-red or bright red; buds depressed; tree.

(C R) OXYDENDRUM

oo Twigs slender, 1.75 mm. or less diam., dull.

p Twigs merely glandular, or glabrous, light brown; shrub.

(C W) ANDROMEDA (*Mariana*)

pp Twigs pubescent, reddish olive; shrub.......... (C W) POLYCODIUM

www.ingramcontent.com/pod-product-compliance
Lightning Source LLC
Chambersburg PA
CBHW070628280225
22716CB00049B/11